BIRDS THAT STAY

POEMS BY SUSAN WATSON

To [illegible] - sister
fellow poet
with love & thanks
for your support -- Susan

ARROW GRAPHICS & TYPOGRAPHY
MISSOULA, MONTANA

Grateful acknowledgement is made to the following publications in which some of these poems have appeared: *Scratchgravel Hills, The Alaskan Quarterly, The Montana Eagle.*

I would like to thank all my friends and fellow poets who have encouraged and supported me in my work, particularly Carol Abrams and Robert Bayuk, who typed the manuscript, Anne Appleby, Melissa Quasny, Rick Newby, Robin Hatch, and Claudia Montagne.

This volume is published by the Montana Arts Council as the winning selection in the 1982 First Book Award Competition. All rights are reserved to the author. Funding for this project was received, in part, from the National Endowment for the Arts.

Printed in the United States
ISBN 0-939872-02-1

Produced by Arrow Graphics & Typography
P.O. Box 8507, Missoula, MT 59807

The Southern Custom—of the Bird—
That ere the Frosts are due—
Accepts a better Latitude—
We—are the Birds—that stay.

Emily Dickinson

For my parents, Anacatherine and Bill.
And for my son, Tygen Absalom.

Contents

I Accepting Poison

II Weather Change

III Out There, It Is Humming

IV Birds

I
ACCEPTING POISON

the common woman is as common
as the common crow.

Judy Grahn

FEELING CHINESE

Everything is burning. Old men
in downtown Noxon stare past the river,
discuss the weather. Everyone agrees:
this season the most beautiful. Earth
as a woman of fashion. She accepts
the end of adolescent attire.
Colors her short hair henna.

As I gather pumpkins before the next
hard freeze, can the last six quarts
of tomatoes, read to my son: *a fruit*
is the ripened ovary of a single flower;
I mourn the passing of huckleberries,
dream of Peru where the Hummingbird
will travel. I remember my childhood
when all dreams were possible.
I study the willow, true to ancestors
who programmed their subtle sad posture.
I am wondering: will I drive suicidal
into the eye of a blizzard another January?
I am feeling Chinese: blue, purple, jade green.
Detached and lost in centuries. Into the wind,
I lift my wood flute. And I hear distant wings.
Fluttering, fluttering. Tiny embryo.

Hydrated Quartz Is Called Moonstone

> The highest good is like water. Water benefits
> all things generously and is without strife.
> It dwells in the lowly places that men disdain.
> Thus it comes nearest to the Tao.
>
> *Tao-te Ching*

I am taking a bath again,
seeking to melt myself in hot water.
Fire sings in the Monarch as I witness
the crackle of dead worms.
I watch my child lick his bowl
of goat yoghurt and strawberries.
Sweet fruit we picked on hands and knees
that hot July day. I'd just come back
from Connecticut and could still hear
my mother scream: *That man has brainwashed you.*
He will never wash dishes. He floats in a dream
you keep feeding. I am feeding the child,
feeding the fire with larch
the man's strong hands have chopped,
and the boy carried in on the broken wheelbarrow.
I am heating water for my bath.
Returning to water, so hot, my woman friend warns:
You'll destroy the collagen in your skin.
Men only love women for beauty. You must watch out
for hot liquids. Even coffee will paint purple lines
on your legs, streak grey in your golden hair.

I forget my fear, wash out nightmare, wood ash
engrained in my fingers; take off my clothes

that don't fit. I don't spend his money on silk,
on adornment. Money he earns in the woods.
I fish my costumes from the dump across the river.
The old man who guards the garbage and leers at me
doesn't suspect I know the secret of his daughter
hiding behind the curtain. I erase what I have seen.
My eyes shut as I sink deep, look at color
filtered through eye lids, color in cells
that are swimming. I seek immersion,
effacement of all trace of blood, bruise,
and stain. Irritation produces pearl.
Hydrated quartz is called moonstone.
I am soaking this in. Skin glistens
in night light. I listen for tide call,
yield to the pull of the moon.

Jesus, If I'd Met You

Jesus, if I'd met you on that pockmarked highway
that crawls through quiet reservation towns
where gardens wave sunflowers; tar paper roofs
signal smoke from aging Majestics; I'd have stopped,
picked you up. Tired, hungover hitchhiker.
I'd let you ride silent and sullen
past mullein and knapweed, the nodding yarrow,
until you could trust me with your story.
Beaten by police in Polson. Stoned
on cactus you'd just swallowed for a vision.
I could believe in such a wish to dream.

And if you'd come running down the Bull River valley,
from Eureka, Clyde Rector after killing a cop
out to kill you; come to my door for shelter,
I'd have harbored you, given you B vitamins and chamomile
to balance your understandable paranoia.

O Jesus, when my teachers told me
I had a vocation, asked me to join them
in devotion to your memory, I prayed
and prayed, asked you simply
to pop the question in person.
But you never came. Not in my bedroom
where I confessed my sins, not in my mouth
where I took you in each morning
in that ancient ritual.

And Jesus, if you just sit dusty
and dry in that worn black book,
I guess I'll never buy your redemption
story. I've already read it, cover
to cover, over and over, until I could quote
any poetry you ever wrote. Jesus, if
you weren't dead, I might love you –
gentle brother, sensitive to the grace
of the lily. I too could kneel,
moved to weeping, as Mary Magdalene wept,
before your kindness, your mystic touch.

NOVEMBER WEATHER

One can not run from bad weather
in Noxon, Montana. It surrounds:
grey, grey-green, blue-grey.
A kingfisher waits by the water,
never wipes rain from feather.
And the rain rolls down the folds
of the body, the fluting of hills.
A wet valley signals danger
around the next unexpected curve.

An edge plunges 300 feet off to my left,
down to the slate-colored river.
Nothing's left, but this getting wet.
Movement of the fluid path of fish,
slippery as sex. Uncertainty,
rain, tears, and waters that run
after a drought that nearly killed
every green and growing thing. Rain
before the final white silence of snow.

Autumn Equinox

Morning of your first winter sickness.
Faded blue flannel and rag wool.
You blend double doses of C crystals,
algae-colored chlorophyll the fat herbalist
ordered from Whitefish. It's too late.
You give in to weather invading your cells.
You want to curl under quilts
and sex does not obsess you.
Tho you'd like a wife in the kitchen –
measuring the medicine, the lemons,
and knapweed honey. Where are the men
whose sore throats and mysterious warts
you healed? Children who call in the night
for a kiss, for comfort, another glass of water?

School's over. Your young son pleads:
I need a knife and a stone for sharpening.
Just like his father's.
The blade precise and cutting clean.

Night and the tiredness. The tyropita baked
to its perfect crust. The firm custard of egg
and cream, right crush of marjoram. You cry
into dishes, then your favorite cup slips from
your grasp like a daughter who plunges,
despite all your adoration, straight to her
smashed ending. Nothing is unimportant.
You are dreaming this death: the smiling rapist
who surprises you, the boss throwing angry rocks.

Father yelling. Mother in the bedroom – devouring
ice-cream. The man you love makes love to a woman:
large, black-haired and quiet. You recognize
your opposite; tell your father, *I'll run away;*
threaten your boss, *I'll form a union;* promise
your lover, *I'm moving to New Hampshire.*
Then you remember the child who dreams of carving
round wooden animals. When you awaken, say to him,
Be careful not to hurt with the sharpness of that knife.

ELEGY

For Burma Rhodes, 1947-1975

I

Under the Dog Star Sirius,
Burma, dance in light. Today, I see you
before you became a bellydance princess:
peeling diaphanous skeins apart like skins,
tapping your hips to air and Turkish music
as life unfolds fast. Your skin turned
translucent as rice paper in China. Eyes
like lakes in Thailand. Burma dancing.

II

Inside Kuskay's sweat lodge,
we shovel burning rock. Canvas tarp
over willow branch. Scent of cedar smoke.
With skin steaming, we crawl outside
to winter, dive deep into Muskrat Creek.
Baptism by fire, ice water: November, 1971.

III

We work by diesel light to unravel
medieval tapestries. They've bound
our bodies for centuries. Self-hate
to our necks. Tears. Learning.
It takes us all winter to translate
the one in German: "In 1432, surgeons
in Persia perform clitoridectomies
on women volunteers." We drip sweat
over cookstoves, chop wood, and share

secret recipes: stuffed heart of deer,
elk roast in tarragon. Exchange
cook books: Female Anatomy Made Simple.

When you leave for sun, saguaro,
and cactus worship, you promise
you'll come back some June
when the bush behind the summer apple
is pregnant with serviceberries.
I shyly hug you, as a mother holds
a favored daughter who will travel
to countries and return
only in letters she sends home.

IV

Burma, what were the words
you fell down to that afternoon
as aneurism exploded, floodgates
let go brainwave to sky, desert,
foreign galaxy? Words hardly matter
when matter turns grey. On the ground
your costume covered by stars,
sequin and pearl — glittering
as sun strikes copper, gold thread,
your elk tooth necklace.

V

Grateful Dead play your favorite music:
Wheel is Turning as Kuskay carries
grey ash back to country where he met you.
A chinook blows warm
wordless dust, tiny paper wings, away.

Pablo Sage
For Maggie Konet

Pablo came kicking with funny large feet
at three thirty-five a.m.
Miniature of Scott with musician's
long thin hands. Tiny fish
slipping from you. Cunt bleeding
bright red and letting go life.

So small, you knew
when the doctors took him
to that stark white world
of scalpels, square incubators.

Now you listen to questions.
Even your husband wonders when
you'll climb out of your bed.

And your breasts are so swollen,
they ache with weight. Milk
you pump and pour out each morning.
Memory of eight hours. His brief love.

You are standing in snow.
Noon in Marlow as you await the body.
Everything is covered
by ice, all this whiteness.

Accepting Poison

I went back. Top of a mountain's carved off.
A clean mastectomy. Grey highway cuts through ground
like logic. They've left that dirt road untouched.
Today, I return, remember October, 1970.
An abandoned cabin on the road to Comet.
Like anxious squirrels smelling the damp
scent of snow, we made it ready for winter.
When Sheriff Paradis arrived, we avoided
questions, poured coffee, let him
feed us stories. Now, one stark room
pulls me in. Voices within a Quaker meeting —
murmuring. Single bed sits in the corner.
Table you hung between beams for my breadmaking.
Blue enamel Majestic rusts where once
I baked stuffed grouse, stirred venison stew.

Through translucent plastic, I trace a worn path
to an outhouse; recall that first winter
when wind carried cold to fifty below and brought me
my first abortion. First time a man's fist
against my face. And rage set white fire falling.
Little deaths stored in quiet catacombs. You watched
as prey raced for a hidden place
past the man's killing, the man's hate.

I dig deep, retrieve old bones, fondle pale grey
textures. Ghosts speak to me. I enter nightmare
and listen: a coyote is screaming. While you
sleep beside your second wife, all night

I translate ancient owl warnings. Once I ignored
the language of elements, of objects.
On the wall, one wildflower cries for color.

I search for fragments to shape some form.
Creosote blackens the floor; soot coats
the wood tones, the windows ashen. Mice have
moved in. Outside, a dead fir shows me
bare bone severed six years ago. You cut it
green. No prayer, no apology. If I said
I were sorry, would it matter? Would it hear?

Waist high in weeds that cover our former garden,
I find the Tansy. Even their intense yellow
fades in fall. We've had the final freeze.
By diesel light, I read how Mohabinees ended
pregnancy; smell bitter tea and gag. Pray to
ancestors this dose is not too strong, then
swallow — accepting poison — knowing the clear
voice of morning will explain everything.

The Reading

You enter her room. It is dark and quiet
like a confessional. In her pale green eyes,
anger burns at your arrogance, your unbelief.
You explain: once you were a keeper of rosaries,
drawn-out virginity, and imagined mortal sins.
Her slender hands cut the cards, turn the first one:

Lovers, locked like twin trees under a sky
of light and color. "Your card's reversed."
Heavens sink in a grey sea. The children
are dead. The lovers are enemies.

The next card is atmosphere.
You want to skip this one, forget
the picture: A woman face down
in dirt. Ten swords through her body.

Then the hermit appears. In one thin hand,
a lamp. It warms him on cold blue mountains.
He spends hours listening, chanting.
His wild hair is like wolfskin: whitegold
and shaggy. She warns, "You might become him."

"But here, your power:" The woman waving wands
stands before patriarchs. The gift of tongues.
One by one, she responds correctly to every riddle.
The patriarchs are stunned. They agree not to rape her,
beat her, or force her to marry their sons.

Yet, one card threatens: One black knight on the hunt.
"You might imagine him a god or a genius, call him
Jesus. You could forget everything. Fall in idolatry."
And he shoots straight through your weak green heart.
A man wanting flesh until wanting turns to dismemberment.

"The last is outcome." A woman weighted. A daughter
of nuns and saints. Scapulars and razor blades
around their necks. She hears priests doling out
her penance: "My child, carry more." The card protests.
Sticks of centuries strapped across her back.
Her legs never breaking the long, slow walk.
Far in the distance: a castle towers.
Clear air. Springs within a garden.

You are remembering a garden.
A garden of gold and green. The fruit
just ripening. The scent of orange
and lemon. She tells you:
"Your final card is promising."

II
WEATHER CHANGE

You see that once there was happiness
This is the way to call it back:
Come back! Come back right away!
I am giving up neatness for you

Laura Jensen
Bad Boats "Happiness"

After The French Movie

Outside your cabin, mountain ash burns to red orange.
An absence of heat turns the garden ochre,
weeds, the pure color of snow at sunset. Curled,
cold animal, you wait for some killing. Loneliness
wants a miracle to alter this arctic wind, thin air.

Only alpine berries left. You hang
in the string hammock Tom brought back from Taxco,
plan to pan gold up Cataract Canyon next weekend,
enough to get you South before November.
If you're lucky, you'll find a few huckleberries
holding on. Tell yourself you'll go back to school.

Harsh blue blankets the valley. Remember
skittish birds we counted on fence posts,
watched them rise up, then disappear like angels?
We could see what was happening again. Frozen ground
thawing to let lupine open over the pale hills.

Now the taste of Canada in the air.
You and I hate September. Children pedalling
to ride the season forever before dissonant bells
ring prison. Walls of dull brick.

After the film you saw last week, you replay
the happy ending. Freeze an image
like a farmwoman desperate to store ripe fruit
in her pantry. Frame of the blonde whirling in sunlight.
She is in love. Nothing else matters.
Her eyes the muted color of wild hyacinth.

HAPPINESS

I shut the door against cold, the empty room,
the bed as it is: unmade and carrying
the faint scent of sex. Look, I've kept
the dark sheets uncovered in the light, the same
as you left them. And with the exact delicacy
with which I touched you, I picture two bodies
sailing on an ocean of tropical birds
and deep coral flowers. On the yellow wall
beautiful women in transparent organdy
walk through dusk, wisteria, and lilac.

Dear friend, I want to be them in their
sexual garden that will bloom forever.

I was afraid of morning with its drizzle
and grey beginning, your nervousness.
Alarms reminding me of parochial school,
factories and prison. Something inescapable
hammering away at my corner of happiness.

I know it was here in this room. I tasted
its distinct flavor on your tongue,
your thin lips searching for wetness
and warmth. I could have told you
This is enough. Now the room wants a person
and heat. I've switched off the pilot,
a ritual I perform with these weather swings.
I'm not ready to remake the bed, change
my favorite sheets, or accept an ending yet.
Tonight, on an aqua couch, I sleep
where there is room enough for one small woman.

Birthday Song

Full moon opens bone white.
Winter on your birthday.
Mid-life arriving like a gift
you never expected. Surprise,
look what the cold has given:
cold and signs of killing.

Alone in your house, the quiet
falls around you like snow.
There is cake and a candle,
card from your mother, the women
who have passed through your life
like all the dead seasons.

I could walk across the frozen water
stretching from your hill to mine,
or use the telephone, if there were words
thin wire could carry from my lips.
Perhaps I could send two lazuli bunting
in a small white box. Then you would know
I know what you are dreaming.

Through iced air that hangs
between us, the silence –
I might hum the E Minor sound
designed to make you weep
for pain shared by those
who reach too far for happiness.

On the night of your birthday,
I could deliver a single flower –
one that blooms after you've forgotten
its color: dark red as a heart.
Blood warm, I'd leave it
breathing a deep jungle odor
through the small crack
under your locked door.

Little Gifts

Downstairs, a woman is crying while her husband watches
television. Thick voices break through my floor
when I would blanket my body in another winter,
quilts with clear symmetrical pattern,
and the white pills you brought back from Butte.

I might pull down an early snow, cover every trace
of today—the shrivelled garden, deer bones scattered
in the back yard. Beside my sick child, I too would give in
to sickness and let it take me to another country,
if not for this army of noises: the wind hammering
windows, the walls of my chest swelling with fists. Upstairs,
a man pounds his favorite photograph: a beautiful woman
smiles, waves through snow, the late hour, and her sadness.

In this cold Victorian house, somewhere an absolute calm
wants to breathe. When it did, I could crawl inside,
wait for this storm to die. Like the aqua rug, lie dead
still. And I too will rest when your gifts no longer speak,
repeat common gossip about a man who filled my life
with objects – abalone boxes, a kettle from Portugal,
and earrings of red stone I could pin through my skin.

Now over weeping, you place sentences like cold plates,
logic any mother might understand about departure,
distance and escape. You still call me your lover,
buy a truck for my son who wants someone to play trucks with,
hand me a grey wool sweater to get me through this winter,

my fear of freezing. You say, as it sags, sacklike
over my body: "You're losing weight; you're doing too much."

In an empty bed, sheets recall bodies,
lovers making believe or making love.
I can hear your bird sounds rising
to a pitch that reminds me of panic.
Here is your scent, thin black hair, odd sock
you left. You are leaving. All the reasons you carve
careful as a poem, do not utter anything else.

In the silence that arrives like a gift
for those who wait for two a.m., your last speech
pales before the door's closing, your footsteps
nervous down the stairs. Words you never spoke.

Weather Change

You climb through yarrow, past spikes of mullein,
pick yellowed leaves for smoke at the summit.
All you hear are noises: large wings flapping.
You picture Portugal, your uncle's windmill
spinning white against indigo sea,
think you can taste the grapes and garlic,
want to be the black-haired woman
weaving purple angora on the veranda.

For three days you've been so sick
you've felt skin dissolving. Now imagine
wind could lift you light over the valley.
Virgin attended by goshawk and raven.

You have come to harvest weeds,
know you will arrange something graceful
with dianthus and lavender you saved from summer.
You let wind take your hair like hands
waiting for you to feel their passion.

Wind pounds so hard, you forget
about Portugal, your alcoholic uncle,
your last lover who never loved you.
Like silver threads, cobwebs trail
into air and disappear.

You begin your run back
like deer in December who dance
after season of blood and bullet.
You no longer remember anything
but the roar of wind in your ear.

III
Out There, It Is Humming

This is the Hour of Lead—
Remembered, if outlived
As Freezing persons, recollect the Snow
First—Chill—then Stupor—then the letting go

Emily Dickinson

Out There, It Is Humming

Inside your kitchen, you move your garden,
sprout alfalfa in a large glass jar.
Outside, marigold and squash blossom shrivel.
Yesterday, you said good bye again,
latticed the gaps moths had made all summer,
rescued the red geranium. The post card winks
beneath the flowers: "Ciao, mia bella,"
turns sour as the temperature
feels less and less Italian.

Every time you gather one more bowl of chard,
you are grateful for each green thing
blooming into the hour of its death.
You wonder how it is that dead things
still drag at your blood, ferret
for one last drop. Meanwhile, the leaves
are speaking, wise mothers who push you
out the door before you lose another season.
They know what you refuse to learn:
how it is they let go their green marriage
to apple, pear and elm, yield to the pull
of ground, slow crawling things and decay.

You grow thin again, a farmer whose crop
was cut back by drought, the indomitable grasshopper,
one unexpected hail in July. Like him, you lean
hard and unromantic into this unloving wind,
know only that cold is not kind to the poor,
or weak, a woman alone when the sun is hidden.

You remember that first winter in Montana:
In the snow, horses kicked against ice,
looking for one small patch of grass.

A pink fire hovers above the ridge edge,
one lingering kiss that will drift
like pale smoke to another country.
You want to be that light, that ephemeral
and unthinking. You are afraid
the voice inside your courage is afraid
to begin again, unwilling to battle
against the wind. Out there, it is humming
reminding you of so many flies
trying to get inside your kitchen.

The stillness of an absence is very quiet.
There is no weeping, no words to toss
across a mountain to move a mountain of rock.
Nothing to be explained. Only one last frame
from that silent era: Someone in the rain.
A crooked smile. An ice cold kiss in the street.

You pick up the litter of toys, pick up
a body of bones and white skin, guard over
your child's rhythmic breathing, imagine
his beautuful dream of leaf-eating dinosaur,
cloud butterfly, and dragon; then work
into another morning: steaming the harvest

for December, preparing the tea an old woman
told you would fix everything that is inside dying.

You have never known faith. That is why
as a child you prayed for miracles,
an apparition to wave through the curtain
and answer riddles you wrote down: why it is
someone somewhere goes on crying
for all the dead children: the five year old girl
in a plastic sack outside Missoula, the boy
stuck in the hay mower near Noxon,
your sister's baby, blue in her crib.
South of here, you picture them, in Guatemala perhaps,
under the bed, an ear tilted to the hot night air,
one heart about to predict: the hard masculine knock.

This Is February

Cold wins in this country.
Towards the end of January, the near-dead
let go, leave those who are still moving
to search alone for love or a chinook.

After eleven winters, in exile
from the humid and peopled cities,
I should be familiar with arctic air
that creeps under every crevice,
crawls over the morning floor,
inhabits my bedroom all night long.

I recognize this imperceptible ache
in my chest, off beat heart tick.
And I recall someone once wrote:
"Angina is a Victorian malady,
usually female." I can hear
one persistent rodent gnaw through my wall,
and that old weak scar I imagined healed.

Now, in my fortress, safe from the intrusion
of strangers and wind, I listen to the snow
and footsteps. I am so slowed by this relentless
winter, your cold departure – violent
as a white-out on the road to Warm Springs.

This is February, that time of year when friends
disappear like men gone to look for gold
in Mexico. Those who stay, learn how to walk
inside a blizzard, how to keep the face from freezing.

THERE IS NO SPRING HERE

There is no spring here.
No resurrection of dead men.
When they disappear in January,
you know they don't come back.
Only this wind returns
like sadness. There is a myth
that one year early in April,
narcissus gleamed white and yellow,
were not chopped down by hail.

There is such deceit in the human tongue,
the weather in Montana. You learn
it is easy to read wrong signals.
Even when cottonwood bud their anxious
yellow-green, you still keep
your pilot on, warn your neighbor
from California, he is planting beans
too soon. You tell him to wait
until it's almost too late to grow
anything but stock for borsch. He points
to the sun and you say: "Tonight,
when the moon is wide, a frost."

You trust no one, not the sun in the sky,
not the lover whose mouth breathes fire
that sparks and thaws your body.
When the weather shifts, you know
everything congeals, turns to ice or stone.

THE MOON IS FLOATING

There is nowhere to swim.
She is cold and alone, almost indifferent
to their canticles, to names they name her:
Mother, Beautiful, Queen of Sky. At night,
sometimes when the air is clear, she hears
the crying of children, the dark sound
of pain and despair. Women weeping
a language she comprehends. And she turns,
as only grey rock can turn in its orbit,
listens to the ebb and flow of suffering.
Tears and the wound of earth opening again
and again. She can do nothing to stop it,
returns to her work of pulling the ocean
back and forth, as though it mattered.
As though salt green water
could cleanse the world of blood.

IV
BIRDS

The bird in the air is fire
its feathers that fall are life
of a green of infinite colors
when it becomes visible among the flames
all the birds of the earth die

Homero Aridjis

The Moon That Awakens You Before Dawn

The moon that awakens you before dawn.
Naked, it glitters through your laced glass.
And a fiery ice beckons you
from dark dreamless country,
soft barbiturate of a sleep where you forget
dreams of a moon, opal beautiful and open.

Round faced mother of the night sky.
Moon that seduces like snow.
No thunder at this hour of secret,
just the white kiss, the lightning look
and pure spark that travels uncluttered
through the blue of another world.

Unlike orange flames of the morning,
this is the cooled heat of the galaxy's passion.
Moon of your oldest memory.
Moon that hangs and tugs at your blood,
your warm still water. When the sea shifts,
when the egg slips down, so quietly no one hears,
only your own trembling body, you will crawl from your cover
of sand and mud, yield to the clear laughter
of that black bird who dives deep for food.
You will float through air, cloudless
and thin, become the woman who swims and swims
into the grey moment before the earth speaks.

Snow

Snow. The night breathes frozen water.
An enigma: The moon with its face of rock
assumes the texture of cloud. Your work
is over, and the pain, the violent blast
from the sun. You know that cold is near
waiting for its moment. Still, on a night
like this, you walk into snow, romantic
and unafraid of the dark, the unfamiliar
sound beyond shadow. Sleepwalker,
you are seeking the pale secret of winter.
Why it is that snow is soft and beautiful.
And why lost men lie down and sleep
in its long white arms. You remember
your mother's goodnight kiss.
How cool against your skin, how fragrant
on her neck, the scent she called:
essence of snow pea or snow drop.
You pictured flowers breaking through ice.

The air is moist fog a child blows back
in a dream where he becomes a snowy owl
about to feather and glide into the night
and a radiant whiteness. Glistening snow
that pours itself into tracks of mud
and blood, repaints the desperate face
of hunger. Snow that closes the door
on your dying garden, the buzzing of flies,
and the dull chant of religious sparrows:
calls forth the shy winter wren,

slate-colored junco, blue canadian jay,
and love — a white seed that drifts
in its lonely pursuit of soil, finds
your face gazing into the geometric shape
and intricate pattern of clear desire.

WINTER, HELENA, MONTANA

In the manner that a mother whispers
round words into darkness
returns the frightened child to dreams
where he wings above clear ice water,
so the snow explains winter.

With the faint descent that never startles
the winter wren or waxwing, snow speaks
like touch and your grandmother's homespun
covering your cold limbs with layers
of wool, affection, and pure color.

When sky loosens its long white hair
you picture your future: an old woman
walking into her one-hundredth winter.
She watches December envelop her body
in a caul of alabaster, observes
how an immense silence overtakes
her footsteps as her pale heart
slows like a frozen river.

These are moments when pain freezes.
And you know you are safe in arctic stillness,
this constant pulse of snow. Winter.
More than half a mile above an ocean.
Where mountains divide a country,
separate people of snow
from those of fog and rain.

Inside the calm that is this weather,
you understand it is always a friend –
quiet and reticent as snow itself –
who travels across the blocks of ice
to knock outside your door, reveal a world
as fragile as crystals holding on to glass.

BIRDS

Love and the arrival of bluebirds
earlier than expected. See, they are nervous.
In their fragile awareness, they know
curiosity is the killer of wild iris and shy creatures
who hide in cedar, nest in alpine meadow.
So vulnerable to the harsh sound,
odd movement at the edge of horizon,
they sometimes forget why it is they are born
to sail blue wind, travel to foreign country.
This is love: we let birds understand our reverence
for all they have become in their lonely flight
toward mountain, their knowledge of time and season,
right moment for departure, and mysterious bird dreams
bringing them back, just as we dreamed them back
in our dreams we share with no one.
We ask nothing of them, but to be here
in this blue hour before dusk
while there is still time to witness the color.

Morning. Peeled wallpaper reveals erotic fantasties:
Blue lilac and vulva-like leaves. Lean and agile,
you stretch and yawn. I watch you, as a mother, in delight,
observes her child's first steps. These days swirl,
as finger against finger, we spin. Two black crows diving
within the sky's current, adjusting our bodies again and again
to each rise and fall of temperature. You are no starling

or robin who sneaks into an orchard and never leaves
once he learns I am a gardener of strawberries and pears.

Love. I keep my distance.
Unlike those whose net and bullet
eventually break everything of color
or wildness, I take my place among weeds,
beside rocks who have studied the art of acceptance
for centuries. And I listen for the sound of wings.
I know it is coming, the many-tongued voice of the birds.

FLOWERS

They are the shape of trumpets
and windmills in Portugal. The color of sky,
of sea, your own blood, and the sun at dusk
and at dawn. All colors you have ever imagined.

What the child of sweetness picks
along the alley for a woman.
She stands at the door, watches for a child
carrying his picture of a house and two tulips.

While a woman takes her bath
alone in late afternoon,
her perfect lover opens the door,
and into the water, drops one blue iris,
kisses her trembling shoulder.

What you wear in your thick black hair,
behind the left ear, when you are wild
and want to dance to drums, Carnival in Rio.

When you are young or in love, you will roll
on a blanket of yellow violet and mountain
forget-me-not. Bold flames of Indian paint brush
swell black and vermillion. And as your eyes shut,
you float calm on a lake of rare golden lotus.

They are the delicate texture of a baby's skin.
Moist in the morning. The inside of a woman
yielding her nectar and fragrance.

And when there is sadness, cold and the rain,
you can gather them to you, fill the empty spaces
with the gladiolas of the grandmothers who remember,
gardenia of the black blues singer you have become,
pale pink rose of the Virgin assumed into heaven.

While they rest beside your bed
in their cool clear water –
lavender lilac and honeysuckle,
you begin your task: to forget,
to forgive, to sleep at last.

And they will heal like a mother
who decorates her Victorian home
with bright zinnia and chrysanthemum
after she has worked with the dirt
growing what is beautiful and green.
Then, as you drift to the lullaby,
they explain their mystery,
their ancient flower secret: how to close
how to open, open then close, and open again.

BIRDS THAT STAY
was printed in a limited edition of 400 copies. It was set in Neufville Schneidler by Arrow Graphics of Missoula, Montana. The papers are Gilcrest Laid Bond and the fly leaf is Ultra II. The book was printed by Gateway Printing of Missoula, Montana.

Cover illustration and design by Allen Woodard.

Charcoal drawings on pages 20 and 30 by Anne Appleby.